NEW LIFE TO THE LAND

By
GEORGE WOODCOCK

FREEDOM PRESS

First Published . . . 1942
Reprinted . . . . August, 1942.

PRINTED IN GREAT BRITAIN
BY EXPRESS PRINTERS, LONDON

INTRODUCTION

During the last two years, and particularly since the war assumed a dynamic character with the defeat of France, the problem of food has attained an importance in the minds of British people greater than that of any other aspect of the war.

Unless we have the money to feed at really expensive restaurants, we find our food lacking in quantity, variety and nutritive value. We cannot obtain enough to eat, and what we get is not sufficiently pleasant to compensate for its deficiency in quantity. Food, like everything in wartime, deteriorates in geometric progression.

Yet, though discontented with the food we have, we are also fearful that as the war progresses we may receive less and again less. At the end of the narrowing perspective of a long war we see starvation and famine, and we are never completely reassured by the propaganda voice of the political boss.

At present England is dependent for a very large proportion of her food on imports from America and the Dominions. Even the scanty rations could not be maintained if it were not for these imports. In the present condition of the agricultural industry in Britain, a long and efficient blockade would mean, literally, famine in this island.

The average man, the man in the street, realizes these facts. But he does not see any way out of them. Whether he is conservative or holds radical views, he accepts without question the current fallacy that it is impossible to grow on British soil sufficient agricultural produce to provide all the population with the essential foodstuffs.

A few public men, including David Lloyd George and Sir R. G. Stapledon, have attempted to destroy this popular fallacy. From intimate knowledge of the potentiality of British agriculture, they realize that the land of Britain has the capacity of producing an abundance of food for every inhabitant of this island. For the most part, however, they are under the misapprehension that the necessary expansion of agriculture can occur under the present economic system, whereas, in fact, the very nature of an imperialist capitalism demands a weak home agriculture to allow for large food imports to balance exports of manufactured goods and secure payment of interest on foreign investments.

To the revolutionary this question of the feeding of Britain is of peculiar importance. For, if adequate food can be produced only after an economic and social revolution, it is equally certain that a revolution cannot be maintained indefinitely unless it secures the provision of adequate food supplies. A country in revolt, even more than a country at war, must provide against a blockade of the most ruthless kind. Revolution without bread is doomed.

The purpose of this pamphlet is to state an anarchist position with regard to British agriculture. After an examination of the decline of farming and the circumstances that have made this inevitable in an imperialist capitalist order, I shall discuss the technical considerations governing the growth of sufficient food in this country, describe the present condition of the farm population, and, finally, outline the form of agricultural organization in an anarchist society and the methods by which the struggle for a better system can be conducted in the country.

It is hoped that these proposals will not be regarded as academic merely. On the contrary, it is desired by the anarchist movement to build a body of active support for its policy, and we shall be pleased to hear from any farmer, farm worker or other person interested in agriculture who cares to write to us.

I

AGRICULTURE BEFORE THE WAR

SIXTY years ago, at the end of the high period of British agriculture, the farming of this country was second in quality to none in Europe. A high proportion of the farm land was under the plough. Crop yields were good (in 1881 Denmark was the only country whose wheat yield was greater than the British), and the cattle of England had no continental peer. In 1870 the bulk of our major food requirements, with the exception of sugar, were still satisfied from English soil.

During the years intervening between 1880 and the beginning of the present war, the position of British farming, both in relation to its own former condition and to the present condition of the advanced agricultural lands on the continent, had declined catastrophically. The areas under cereals and root crops had decreased by nearly 50% since 1870. The yield per acre of English wheat, while it had increased to a small extent, had become fifth among western European countries, and was below that of Denmark and Holland by more than 25%. Our cattle had fallen in quality far behind those of Denmark and Holland, whose milk yields were now 40% higher than those of English cattle. While since 1880 the production of cereals had increased in every Western European country (in Denmark by 60%), in Great Britain it had fallen by more than a quarter. At the same time, the increase in livestock population was only 11%, as against 134% in Denmark, 74% in Holland, and 43% in Switzerland.

Imports of food had increased phenomenally. If we compare the periods 1861-5 and 1932-6, we find that in the latter period annual imports of grain were more than three times, sugar nearly three times, cheese four times, butter eight times, and *meat twenty times as great as in the former period.*

Thus, in a period of increasing population and increasing individual consumption of food, a period during which agriculture in every other country of the civilized world had advanced steadily in production and technique, British farming stood still and even, in certain major directions, retreated from its former position. From being the major industry of the country, producing the bulk of the essential foods consumed by the population, it became an industry devoted for the most part to the production of those more-or-less perishable goods which could not conveniently be imported from abroad.

To determine the reason for this decline in the farming industry in Britain it is necessary to return beyond the period of evident decay and trace the way in which the economic and political development of British capitalism made inevitable such a decline.

The agricultural revolution, which inaugurated the heroic period of British farming, began in the years following the Civil War. The Restoration atmosphere of scientific speculation encouraged a series of technical experiments, by such cultured landowners as Temple and the famous Evelyn, and from the resultant improvements the production of food began to increase to such an extent that for a century British farming passed from a subsistence to an exporting basis. The landowning class encouraged this tendency by granting in 1685 a bounty of 5s. a quarter on exported wheat, and by the middle of the eighteenth century the trade in wheat was considerable (9,515,000 quarters were exported in the twenty years 1746-65). There was also a large export trade in malt and barley, and in England itself, and especially the expanding city of London, there was a steadily increasing demand for farm goods.

Under the stimulus of this growing market, improvements in technique were continuous, and cumulative in their effects. Turnips and clover came into general use, and precipitated the transition from the old and wasteful three-year rotation to the more efficient four-year rotation. Improved implements made possible deep ploughing and cultivation. The introduction of root crops caused a revolution in the livestock industry, making it possible for cattle to be fed adequately during the winter and ending the system of slaughtering and salting the majority of the cattle every autumn because of the lack of fodder to maintain them. The improvements in feeding and breeding technique had such effect that, while in 1710 the average weights of sheep and cattle sold at Smithfield were 28 lbs. and 370 lbs. respectively, in 1795 they were 80 lbs. and 800 lbs. respectively. As a further instance of the cumulative effects of improvements in agricultural practice, the new system of stockbreeding, made possible by the introduction into arable farming of the turnip, in its turn assisted the further development of arable farming by making available large supplies of manure and enabling more intensive fertilization.

These improvements in agricultural practice were mostly due to a number of progressive landowners, such as Jethro Tull, 'Turnip' Townsend and Coke of Holkham. Indeed, it could hardly have been otherwise, for the experiments necessitated large outlays of capital and large compact areas of farmland, as opposed to the open-field strip farming and small-scale yeoman farming that still covered large portions of the country. The gentry acquired a vested interest in the improvement of farming, and this led to important social changes in the rural areas. The landowners became greedy for more land and, as they were the dominant class in both Houses of Parliament, they used the instruments of legislation to satisfy their land hunger, and justify their appropriation of the commons. The peasant class was swept from English history, the small yeoman farmer was squeezed almost to extinction, and the social structure of English farming became that of large-scale tenant farming, varied by direct farming on the part of the more enterprising landlords. This system persists to the present day, when little more than 30% of English agricultural land is farmed by its owners, as against 95% in Denmark.

It is impossible to say what might have been the social and economic development of England if it had remained a mainly agricultural country dominated by the landowning aristocracy and gentry. But the nature of

British economy was changed and the power of the landowners broken by the rise of capitalist industry which commenced during the latter part of the eighteenth century and was accelerated by the Napoleonic wars. By a curious irony, the expansion of industry was assisted by the improved methods of farming, which threw out of work many of the rural workers and so created a labour pool to satisfy the needs of the new factories.

The industrial revolution created a moneyed class who realized the necessity for political power, if their interests were to be maintained, and who were debarred from this power by the laws that granted parliamentary suffrage only to landowners. A bitter struggle ensued, in which the industrialists recruited the support of the working classes and achieved final success in the Reform Bill of 1832.

The ascendancy of industrial capitalism marked a change in the attitude of the governing class towards agriculture. It had been in the interests of the landowners to restrict imports of food in order to keep high the prices they could obtain for farm produce, or the rents they could squeeze from the tenant farmers. To ensure the maintenance of price levels they enacted in 1815 the Corn Law which imposed a high tariff on imported corn. But it was in the interests of the industrial class to have plentiful supplies of cheap food so that they could reduce labour costs and so compete more effectively in the world markets. In 1846, after a seven-years' struggle, the Corn Laws were repealed, and the great era of free trade commenced.

Free trade meant in practice the freedom of British manufacturers to sell their goods in the expanding markets of undeveloped countries, where they had few competitors. But they had in some way or another to receive commodities in exchange for their exported manufactured goods and as interest on the surplus capital they invested abroad. So with the export trade in manufactured goods grew up the parallel import trade in food and raw materials. The basis of English industrial capitalism thus became the balance of exported manufactures and imported food, and it is the vital necessity of preserving this balance that has dominated to this day the policy of the British ruling class towards agriculture.

There was a time lag from the repeal of the Corn Laws during which the imports from undeveloped countries were still small and British agriculture continued to develop to serve the expanding home market. Then, from the 1860's, the competition from foreign sources became so great as first to satisfy the increase in demands and then to attack the portion served by British agriculture. For a while home farming remained stable but stagnant and then, about 1880, the decline began before the ever-increasing imports, first of grain, then of meat, butter and cheese, both from the expanding countries of the New World and from the countries of Europe, such as Denmark, Holland and Switzerland, whose agriculture was growing to meet the increased foreign markets.

During the years between 1880 and 1914, British farming ceased to be a major industry growing the bulk of the nation's essential food, and became instead an industry based on those food demands, such as liquid milk, green vegetables, etc., which could not conveniently be met by imports from the dependent countries.

NEW LIFE TO THE LAND

The war of 1914-18 placed the capitalist class in a dilemma regarding agriculture. The blockade made it necessary for more of the essential foods to be grown at home. Indeed, it would have served the temporary purpose of the British rulers if for the duration all the essential foods could have been reared in England. But they desired to return after peace to the old system of trading and realized that this would be incompatible with a home agriculture on a permanently self-sufficient basis. So they resorted to various unsatisfactory measures which had to counterbalance the damage done to agriculture early in the war by the immoderate calling up of farm labourers, the commandeering of farm horses by the army and the scarcity of the means of repairing farm machinery. In spite of the manifest difficulties, an appreciable increase in production *was* achieved. Some two million acres of land were put under the plough, and the combined yield of grains, peas, beans and potatoes, which in 1914 was 14 million tons and in 1916 as low as 11½ million tons, rose in 1918 to 18 million tons. Farmers enjoyed temporary prosperity and the few remaining labourers an equally temporary improvement in wages.

After the war the trend was reversed. Government support was withdrawn from the farmers, who soon found it difficult to make ends meet, and had to cut down farm staffs and other expenses. The production of corn sank to a new low level, and those farmers who were situated favourably became specialists in dairy farming, poultry rearing and market gardening.

The British governments in the period between the two wars were faced with the problem of maintaining at least a skeleton agricultural industry, which could be expanded in time of war, while not allowing it to compete at all seriously with the peace-time imports of food necessary to preserve the capitalist economic balance. Various schemes were attempted, but in general the policy followed was that of issuing subsidies and at the same time limiting production to such a degree that a substantial field remained for the imported product. Certain measures of rationalization were introduced, resulting in the erection of a bureaucratic framework, in the form of marketing boards, about the principal branches of agricultural production, which was expensive and of little value to the farmers or the consuming public but would form the nucleus for a system of control during wartime.

Thus in this period the capitalist state showed its complete social worthlessness in attempting to solve its agricultural problems by restricting the production of food in a country where, through the economic depression, a very large proportion of the people lived well below the level of nutritional sufficiency.[1]

[1] See Sir John Boyd Orr *Food, Health and Income, 1936.* Mc Gonigle and Kirby *Poverty and Public Health, 1936,* etc.

2

THE GOVERNMENT AND WARTIME AGRICULTURE

THE onset of the war in 1939 precipitated no fundamental change in the Government's agricultural policy. The dilemma of the needs of capitalist economy opposed to the needs of a hungry country in wartime remained, except that the war was now in the present and not in some problematical future. But there was a superficial change, in that the emphasis was shifted from one horn of the dilemma to the next. The pre-war problem of the British Government was to keep an extensive market open for imported food while preserving from decay a skeleton of farming that would provide the basis for producing in wartime a large proportion of the nation's food. The wartime problem was to produce a large proportion of the nation's food while making sure that after the war it would be possible once again to open a substantial market for foreign food.

It has long been pretended by capitalist and Marxist economists and believed by the majority of the general public that it is not possible to grow on British soil sufficient food to support the present population of the island. In the next section of this essay we shall examine this fallacious idea. For the moment we shall content ourselves with asserting that, while the spokesmen of the ruling classes say that *sufficient food cannot be grown at home*, the truth is that *they do not want to grow sufficient food at home*. It would create a precedent which, if it caused the elimination in peacetime of the bulk of food imports into this country, might impair the export trade in manufactured goods which they hope to regain at the end of the war—at the expense, of course, of their defeated imperialist rivals.

The existence of this motive among the ruling class is demonstrated by the following quotation from the semi-official organ, *The Times*, which appeared on 11th January, 1941, more than a year after the commencement of the war and more than six months after the military collapse of France:

> 'Any further decline in British imports of food would be little short of a disaster, for the prosperity of British industry and British shipping has always depended, and must continue to depend, largely on the fact that Great Britain is the richest market in the world for the exports of predominantly agricultural countries.'

This statement does, I think, represent fairly the attitude of the ruling class and the bureaucracy towards the problem of wartime food production in Britain. That such an attitude is maintained by the Government is shown by official statements that even in wartime 25% of our food must be supplied from America—to which one must, presumably, add the supplies obtained from Canada, Australia, etc. It would not be unreasonable to suppose that from all sources the government anticipates importing during the war at least 40% of our food requirements, and that they do not hope—or intend—to grow more than 60% on British soil. Such a percentage of wartime production would represent a considerably lower proportion of peacetime consumption. So that, even if the increased home production of food were maintained for some time after peace, the cessation of hostilities would probably find nearly half the British market open to imported foods. A return to a policy of restricted production would further enlarge the market for foreign food. What the capitalists do not consider seriously is the possibility that after the war the world may not provide even the relatively restricted market for British exports which it did in 1939.

It is in view of these considerations that we must examine the somewhat inadequate organization that the government has provided for increasing the production of food in this country during wartime.

To an extent this has consisted of an extension of the compulsory marketing schemes imposed before the war. To the marketing regulations for such staple products as milk and bacon have been added others covering a much wider field of agricultural produce, down to such crops as onions and carrots, until now there is hardly a single agricultural product of any importance whose sale is not so governed.

Marketing must be considered in connection with the various food rationing schemes, and also with the regulations for fixing maximum prices. In general, the effect has been far from good. Both before and during the war the machinery has imposed a heavy burden on the industry. The milk marketing scheme was always a subject of grievance among farmers, and resulted in no appreciable advantage to consumers. The same applies to the wartime schemes, many of which have proved extremely wasteful in valuable food owing to the inefficiency of administration.

The egg rationing scheme took away eggs from areas where they had been plentiful, but no really adequate supplies appeared where they had been scarce. The total supply of eggs was reduced, because to many of the smaller poultry keepers who had sold their eggs retail the prices offered by the Government did not afford a profitable margin. They therefore killed their poultry.

In the case of onions, there was much wastage owing to deterioration. Growers were forced to sell to a government marketing company, and there have been cases where collection has been so late that the major part of the onions have rotted before they reached market, without the grower being able to sell them to prevent deterioration. One grower lost more than 60% of his crop for this reason. Among the anomalies of this scheme is the fact that where growers sell their produce in their own shops, they are not allowed to retain their quota of onions, but have to send their entire crop to the marketing company and are then returned the portion the authorities graciously allow them to sell. Unnecessary transport charges are thus involved, and from this superfluous transaction the marketing company rakes off at the expense of the grower.

Price control schemes have been unfair both to the consuming public and the small grower. In the case of tomatoes, control was not imposed until the large capitalist growers with extensive hothouses had enjoyed the benefit of the enhanced prices. When the smaller growers entered the market they received prices which were fractions of those operating early in the season. Tomatoes were sold retail at 1s. 1d. per lb., and the maximum wholesale price was 9⅓d., but I know from my own experience that the producer by no means always received the full wholesale price—if he sent to market he was mulcted for high sales and transport charges which often reduced his taking to less than half the retail price. In the case of apples, although poor quality fruit were seldom sold in the shops at prices below the maximum, the small growers rarely obtained the maximum wholesale price for any but the best fruit. Grapes, mushrooms and other crops grown mostly by large capitalist firms have never been subjected to control, with the result that they yield exorbitant profits to the few growers and the ubiquitous middlemen.

But, although these forms of organization for food distribution are worthy of consideration in that they retard production, increase costs and reduce returns to all but the large capitalist growers, they have not the basic importance of the experiments the Government has initiated in food production. I use advisedly the word *experiments*, because the Government's actions in this direction, as in many others, must be regarded as tentative rather than based on full and mature deliberation.

The most important of the wartime organizations for food production are the County War Agricultural Committees. Theoretically, these committees administer the whole of local food production, including dictating to the farmers what crops they shall grow on every acre of the farms, the ploughing of fresh land, livestock rationing, the provision on credit of tractors, fertilizers and seeds, and the extermination of pests. They have power to take over and farm land that has been neglected, and employ a considerable pool of labour, including conscientious objectors, land girls, and Italian prisoners, for farming these lands and for assisting any farmer who finds difficulty in obtaining the workers he needs.

Ostensibly, the committees consist of representatives of all three interests directly concerned in farming, the landowners, the farmers and the land-workers. In practice, the landowners and the very large farmers make the decisions. Small farmers, smallholders, crofters and market gardeners are virtually without representation, and the farm workers' representation is often nominal. On a committee for which I worked as a labourer the 'representative' of the agricultural workers was a permanent official of a general union in which only a minority of the members were land workers (the T. & G.W.U.). He himself had never worked on a farm, and was so reactionary and so fearful of his colleagues on the committee that on one occasion, when a dispute arose regarding the working conditions of one of the mobile gangs, he excused his failure to defend their case by saying that 'the man sitting next to him did not like conscientious objectors.'

The decisions of each committee are administered by a number of permanent officials. On the committee I have mentioned none of the responsible officials were men of any technical training. Nor were they good farmers, for at the present time there is more money to be made from farming than from a salaried position with a committee, and, in consequence, the officials who had been farmers were men who had failed in their enterprises or in some other way proved themselves unfitted for the direction of a farm. Their attitude to the general question of agriculture is reflected in the advice given by one of them to a friend. 'Put as little as you can into your land, get as much as you can out of it, and sell out as soon as the war's over.'

The Labour Officer, who had apparently no previous experience of labour questions, handled the workers under him with an amazing lack of tact and a brutal disregard for their convenience or welfare. He attempted to impose an almost military discipline on the 600 men and women in his control, and with the majority he was successful. He did, however, become involved in a series of disputes with the mobile gangs, consisting mainly of somewhat militant C.O.'s, which resulted in the numbers in these gangs falling in eight months from 120 to less than 50. Although the trades union was represented on the committee, he refused to recognize it in practice, and victimized the field stewards appointed by the men.

During his term of office the committee on one occasion locked out the majority of the men in certain gangs because they would not agree to a change in the conditions of employment with regard to sick pay, which the committee had granted before and now wished to withdraw. The trade union supported the committee in what amounted to a breach of the Defence Regulation and would grant the men no dispute pay during their time out of employment, on the patently ridiculous ground that they were technically on strike. In consequence, the men had to accept the best terms they could obtain, which amounted to a promise by the committee that they would consider sympathetically each case of sickness and give adequate benefit according to its individual merits. This promise was carried out quite fairly for the first few weeks. Then the benefits began to fall in scale until, three or four months afterwards, many cases were given nothing at all, because, it was stated, single men earning 48s. a week should be able to save enough to provide for times of illness.

The committee conducted a hostel for some of the men in its employment, managed by an ex-sergeant-major, who used barracks methods and imposed many unpleasant restrictions, including the locking of doors at 11 p.m. The men slept on mattresses laid on the floors in crowded and almost unfurnished dormitories, and the food was bad. Men who expressed radical views of any kind were threatened with expulsion, and an attempt was made to forbid the importation of 'subversive' literature, which culminated in a storm over a copy of Herbert Read's *Poetry and Anarchism*! These conditions caused the departure of so many men from the hostel that it was eventually abandoned because of an insufficiency of occupants.

The inability to deal satisfactorily with labour seems to be common to a number of committees, as the Government found it necessary to impose an order forbidding workers to leave the employment of the War Agricultural Committees.

On the committee by which I was employed the actual field work was little better administered than the labour questions. Drains and ditches were dug by methods of guesswork, without any previous surveys having been made—often with the consequence that the work had to be duplicated until the field was a network of intersecting drains. On one field of very poor land no less than £500 was spent on drains which could have been done for a third the price if only a surveyor had levelled the ground previously. On another occasion a series of drains was dug across a field of growing corn. In the process at least an acre of the crop was destroyed. Extravagance was everywhere in evidence. As one of the foremen put it, 'We don't care a b——r about the work so long as we spend money.' On no occasion during my five months' work did I see a surveyor or, indeed, any technical expert come on the field.

I imagine this committee is a bad example, but few of the committees are free from obvious faults. Particularly, they pay too little attention to the small grower, the market gardener and the smallholder, who are rarely represented on the committees. Even in the market gardening Home Counties there are committees which have no horticultural advisers or sub-committees, and therefore little direct contact with horticultural matters, with the consequence that many unjust and uneconomic orders have been issued to market gardeners. Growers were ordered to plant with potatoes or cereals land on which they had grown by intensive methods two or three crops of garden produce a year. Some committees forbade the growth of lettuce (on the ground that it is a 'luxury' crop) and others have ordered the growth of tomatoes in houses suitable only for cucumbers. It is difficult to see what object the committees hope to gain by giving orders to growers which must obviously reduce both the bulk and the cash return of their crops.

Another matter, on which the committees are dangerous in intent rather than through inefficiency, is their co-operation with the new vested interests that are acquiring agricultural land both as a safe investment and as a means of avoiding the payment of profit and income taxes. In one case in the Eastern counties a committee appropriated a derelict farm, performed the major part of the reclamation work with their own labour, and then transferred the farm to a large canning firm which is acquiring large areas of agricultural land in the district.

Nevertheless, some of the committees, within their unavoidable limitations, work efficiently and are of assistance to the farmers. In Essex, for instance, a community farm on which I worked was given useful advice and help by the Committee's representatives. And in Devonshire a number of tractor stations have been organized to serve small farmers.

Taken as a whole, and with due regard to the difficulties under which they work, the committees have achieved much that could not have been done without organization. Up to last harvest they caused some two million acres to be put under the plough and they have reclaimed many derelict farms and caused many backward farmers to adopt more modern methods of farming.

It should, however, be noted that the two million acres has by no means entirely been devoted to the growing of human food. A very large proportion was used for growing animal foods which often did not leave the farm on which they grew, but were used by the farmer to maintain his existing herds, whose allowances had been curtailed by the rationing authority.

And, whatever the achievements of the committees, no function they perform could not be done better by a collective organization among the farmers themselves, which had in view not a temporary wartime increase in production, but a policy of co-operation aimed at the feeding of the British people from their own soil. This would eliminate the possibility of a dictated routine crushing out initiative among the farming classes and would also offset the danger that exists at the moment of the Executive Committees becoming the instruments of the present sinister trend among industrial and financial groups to create a new vested interest in agriculture, which would certainly be beneficial neither to the farming population nor to the consumers of the food they grow.

3

FEEDING THE PEOPLE OF BRITAIN

As I have already pointed out, the agricultural policy of British governments (in so far as one can term a policy the series of expedients which followed the realization during the thirties that farming was a moribund industry) has always been founded on the assumption that it is impossible to feed the people of Britain from the soil of these islands, and that the most that can be done is to maintain home agriculture by subsidies of one kind or another, to enable it to reduce substantially the proportion of food imported in wartime. This attitude is a natural development of the general attitude of imperialist capitalism, which depends for its continuance on large imports of food and other consumption goods to balance the exports of manufactured goods, without which, under the capitalist system, industrial concerns would not survive. It is not, therefore, surprising that it should be accepted by the majority of economists. In such a fabulous task as the justification of capitalist economy, any fiction can be maintained.

Nor is it only the capitalist economists who support this fallacious point of view. The theoreticians of the 'Left' parties, frozen in their beatific contemplation of a dictatorship of the industrial proletariat, and concerned in their immediate policies almost exclusively with the problems of the industrial worker, have given little consideration to land organization and the growing of a sufficiency of food in this country. Examine a typical Labour Party, I.L.P. or Communist Party programme, and you will find that, except for a token demand for nationalization of the land and some half-hearted proposals for improving the status of the agricultural labourer, there is little in the way of a constructive and truly socialist agricultural policy that aims at making full use of the productive capacity of British soil.

Generally speaking, the Left politicians fail to realize the revolutionary importance of an agricultural policy that will ensure adequate food without the need of supplies from overseas. They appear to act on the assumption that, if the capitalist state were overthrown in this country, food could still be obtained from other capitalist countries. Alternatively, they imagine their revolution taking place simultaneously with those of other countries in the great and millenial world revolution.

The world revolution of this kind is an old, but elusive, friend, and so unlikely that we need not give it serious consideration. As for the idea that capitalist countries are likely to feed a country in social revolution, the reverse is almost certainly true. The great problem of the British revolution, whatever form it may take and whoever may lead it, will be the problem of bread, of feeding the people as far as possible from the sources immediately at our disposal. Unless the importance of this is realized from the beginning, unless an adequate policy is worked out that will make unlikely a starvation of the people by means of an economic blockade, the revolution will certainly fail.

Yet, in spite of the particular importance of food, the soi-disant revolutionary socialist parties have given it little attention. In general, they are content to accept without criticism the ideas of the capitalist economists of the impossibility of growing on British soil all the necessary foodstuffs, and so little real attention has been given to agriculture in radical circles that even Stalinist Communists have admitted to me that there has been no constructive survey of British agriculture since Kropotkin.

There are two further important reasons for favouring the feeding of the people of Britain from the soil of the country. Firstly, the food would be more nourishing and more palatable because, if an efficient and speedy distribution were arranged, it would not have to be subjected to the various preservation processes which lower the value of so much food under the import system. Secondly, if agriculture became once again our major industry, the country would regain its former importance in national life, and the growing flow of population back from the cities would establish a new contact between the rural and urban areas. Thus the town masses would be brought into touch with a healthier manner of life, and the present unhealthy preponderance of the city aggregation could be broken down. An illustration of the benefits which might accrue from such a development can be gained by comparing the mental vitality of the Welsh and English miners, whose conditions of life allow a close contact with the country, with that of industrial workers in the large urban districts.

In fact, as I shall endeavour to demonstrate, the feeding of Britain from its own soil is an object not so distant as is generally imagined.

Just previous to the war, we produced at home 25% of wheat consumed, 55% of barley, 92% of oats, practically all our roots and potatoes, and beet sufficient for 30% of our sugar. This includes the feeding stuffs necessary for cattle bred in this country, but does not, of course, account for meat and dairy products imported. The home production of meat, including bacon and poultry, was approximately 55% of our consumption. The home production of eggs was officially estimated at 64%, but was probably higher, as at that time there was no efficient means of keeping a check on small producers and backyard chicken rearers.

But from 1870 to 1937 the area under corn crops had fallen from 9½ million to 5 million acres, and that under roots from 3⅔ million to 2 million acres. If we returned to the arable acreages of the peak period of English agriculture, we could (assuming yields on the same scale as at present) produce all our barley, oats, root crops and potatoes, more than half our wheat, and probably the whole of our sugar.

Furthermore, in England only a very small proportion of the land is cultivated to its maximum capacity. For the last half-century agricultural technique has moved sluggishly in comparison with the rapid development in the Continental exporting countries. Agricultural research has proceeded in the colleges and research stations, but as British farming has been a diminishing rather than an expanding industry it has had no room for the discoveries that would result in notably increased production, and in consequence agricultural science in England has become more and more an academic science divorced from farming by the economic situation.

The nature of British farming, with its large tenant farms and relatively low rents, is such that a satisfactory financial return can be obtained from a farm whose actual yield per acre is low. The value of agricultural produce in England relative to the area of farm land is the lowest in Western Europe, the net figure being £6 per acre as against £14 in Holland and £15 in Switzerland. There are a number of important reasons for this fact. In all the other countries more labour is expended on cultivation—there are 3 workers per 100 acres of farm land in England, as against 6 in Denmark, whose figure is the lowest on the continent. There are fewer livestock in England—30 units per 100 acres as against 53 in Denmark and 61 in Switzerland, and consequently less animal manure. Artificial fertilizers are used more sparingly—the Dutch, Flemish and Danish peasants use more than twice as much as the English farmers. And mechanization, which might have compensated to some extent for these deficiencies, has progressed very slowly.

Under research conditions yields have been obtained which make English crops appear very small. 249 bushels of oats have been grown to the acre, as against our average of 44 bushels, and 80 tons of potatoes against 6½ tons. But it must be admitted that one cannot expect such yields in ordinary farming, and more just comparisons can be made with average yields in such countries as Denmark, Belgium and Holland, on soils and in climates no better and often worse than the British, but with good seed and the plentiful use of fertilizers.

Sixty years ago the Danish yields were no higher than ours. Now their yields of cereals are from 25% to 50% higher (wheat 25 cwt. against 18 cwt.), sugar beet 50% higher and swedes 60% higher. In Belgium the average yields of oats and barley are 30 and 33 cwt. respectively, against 23 cwt. and 19 cwt. respectively in England. In Holland the sugar beet yield is 70% higher than the English, being 14½ tons against 8½ tons, and the wheat and barley yields are 24½ cwt. and 23½ cwt. against 18 cwt. and 19 cwt. respectively.

It therefore appears that, by the standards of cultivation in Denmark, Holland and Belgium, more than a third more cereals and a half more root crops could be grown per acre of arable land. If we were to combine these yields with the 1870 arable acreages, there can be no doubt of the possibility of growing in this country sufficient of the essential vegetable foodstuffs, including sugar and wheat, needed to support the people of the country, with a surplus equivalent to the amount of root crops and grain, both home grown and imported, previously fed to British livestock.

It may be objected that a considerable increase in arable land would mean a decrease in the area of grassland and a consequent decrease in the number of livestock it is possible to raise in this country. Against this objection I would place the following facts.

During the present century arable land has steadily given place to permanent pasture, permanent pasture to rough grazing, and rough grazing to waste land. From 1891 to 1937 the area of ground under crops and permanent grass fell by more than 3½ million acres. On this approximately half a million acres were lost to the unrestricted building schemes of pre-war days. The rest, over 3 million acres, became rough grazing. This, if it were reclaimed, would give half the acreage taken away from grasslands in order to return to the 1870 acreage of arable land. In addition, hundreds of thousands of acres of valuable hillside pasture and rough grazing in Scotland and Wales, and to a less extent in England, have become useless through the spread of bracken and the ravages of rabbits. A reclamation of this land and the scandalous deer forests that absorb so many hundreds of square miles of the Highlands would provide much useful grazing land.

A survey of the grasslands of Wales conducted under Sir R. G. Stapledon led to the conclusion that by intelligent reclamation the Welsh pastures could maintain more than twice their present livestock population.

Indeed, the crop of grass in this country could be greatly increased if it were treated as a crop rather than as a chance blessing from the Almighty. The researches of such experts as Stapledon have shown that pasture and hay crops can be made to give much higher yields of animal foodstuffs by manuring with artificial fertilizers and by careful selection and blending of the grasses and clovers.

Results obtained in Switzerland and Denmark by intensive manuring of grass are here of interest, and I show in the following table the average figures for these countries, compared with the British figures. It will be observed that in Switzerland *two* crops of hay are obtained by cutting the grass when it is short and green, and using it in silage.

	Britain	*Denmark*	*Switzerland*
Seeds hay	28 cwt.	44 cwt.	47+24 cwt.
Meadow hay	20 cwt.	32 cwt.	41+20 cwt.

The increased growth of root crops and the greater output of pulp from sugar factories would mean a greater supply of foodstuffs other than grass.

And here it may be noted that the Danes have effected in fifty years a 134% increase in livestock population (excluding poultry) while the proportion of arable land remains as high as 83%. Furthermore in peace-time English stock were already fed to a very large extent on imported feeding material, which would be replaced and augmented by an increase in arable land. More arable land means, indeed, more rather than fewer cattle.

Taking these factors into consideration, it would seem that, with the reclamation of waste land and rough grazing and a scientific treatment of grass crops, a much greater yield of forage per acre of non-arable land could be obtained. Thus, far from being decreased, the livestock population could be increased at the same time as a substantial increase in the area of ground under crop.

Not only could the numbers of livestock be increased at the same time as a vast increase in arable crops, but there could be great improvements, through breeding and feeding, in the quality and performance of such animals. In England the average milk yield per cow is 540 gallons per annum. In Denmark it is 700 gallons, in Holland 770 gallons—over 40% higher. Under the English system of pig breeding it takes eight to ten months for a pig to reach 180 lbs., while under the Danish system this weight is attained in six months.

Furthermore, a closer attention to animal diseases, which are estimated at present to cause a loss to the livestock industry of nearly 25% the total output, would without doubt result in a further increase in animal products.

It is certain that by a more intensive and scientific cultivation of the land, by the adequate use of fertilizers on both arable and grass lands, by the use of better seeds, by a closer attention to the breeding, feeding and diseases of animals, and by the increased mechanization of farming, British agriculture could be made sufficiently fruitful to grow an abundance of our essential foods.

But before such changes in agriculture can be achieved, there will be necessary a radical change in the economic and social structure of the ndustry. This will be discussed in the remaining sections of this essay.

4

WORKERS IN AGRICULTURE

MANY social historians, including Marxists, regard the Agricultural Revolution as a change that gave a capitalist structure to agriculture and marked the beginning of the rise of modern large-scale capitalism. In fact, the social shift of the Agricultural Revolution was retrogressive, and amounted to a revival of the apparently moribund feudalism, rising from its decline in a new form adapted to the age, as the Holy Catholic Church rose again under the missionary banner of Loyola. The independent yeomen farmers were rapidly squeezed from the holdings and they, with the peasant strip farmers of the common lands, were thrown on the rural labour market. The land came once again almost entirely under the rule of the gentry, and the old feudal hierarchy, based on service, of king, lord, knight and villein was replaced by the new feudal hierarchy (based on cash and stripped of the moral sanctions that mitigated the old feudalism) of landlord, tenant farmer and labourer, the functions of king being performed by the landed oligarchy in Parliament, which granted rights over the common lands just as the Norman kings had granted rights over the lands of the defeated Saxons. This neo-feudal structure of the English countryside has persisted and to-day only 30% of the agricultural land is farmed by its owners—the lowest proportion in all Europe, over most of which feudalism has been superseded by a peasant economy. (In one country only has agriculture become capitalist, and that is Russia, where collectives have been organized from above in the service of the peculiar form of state capitalism favoured by the Kremlin.)

The power of the new feudalism was short-lived and, indeed, died of its own hands, for it was the labour released by the Agricultural Revolution that manned the new factories and gave power to the capitalist industrial magnates. Thus, though the Agricultural Revolution was in itself not capitalist, but neo-feudal, it prepared, by its effects on the rural labour situation, the advent of a capitalist social and economic structure.

Henceforward, the new feudalism of the countryside became subordinate to the capitalism of the towns. Externally, in relation to capitalism, it had to live by the capitalist economic law. But internally it was still based on pre-capitalist standards. From this fact arises the opposition of interests between town and country, between the industrial class desirous of developing foreign markets for manufactured goods at the expense of home markets for agricultural products, and the agricultural class desirous of developing home markets for agricultural goods at the expense—if necessary—of foreign markets for manufactured goods.

Thus, as in other respects, so in its labour relations is farming pre-capitalist, of the age before the flood of the industrial revolution. It is based on the system (which existed up to the end of the eighteenth century over the majority of industry) of the small master, and in considering the problems of agricultural labour we must not fail to recognize this fact and its implications.

In farming the employing class bears a substantial ratio to the employed. All told, there are some 370,000 farms and holdings of one kind and another in this country. There are approximately 800,000 regular agricultural workers, so that it will be seen that there is approximately one master to every two men. Almost all these farmers take an active part in the work of their farms. Many of them employ no labour—some 60,000 of the holdings are below 5 acres, and 165,000 between 5 and 50 acres.

Those farmers who employ labour are as guilty of exploitation as the large capitalists; many are even harder task-masters. But in justice it must be remembered that they too are exploited, by landlords, tithe-holders, middlemen and government marketing boards, and they have to bear the losses of economic depressions and bad seasons. A survey of the Ministry of Agriculture showed that in 1936 and 1937, typical pre-war years, the 100-acre farm (this is about the average size) gave a return of £140—*including the value of produce consumed by the farmer and his family.* In other words, *the majority of farmers before the war received an income no larger than that of a London navvy.* From employers whose economic status is so insecure and whose margin of profit so slight, it is foolish to expect any appreciable amelioration of labour conditions.

It is true that, for the time being, farmers are making good money, but they know very well that once the war is over their position may deteriorate. Farmers in general are cynical regarding the intentions of the Government and fully expect that when farming has served its wartime purpose it will again become one of the neglected industries, in which earning a living will be difficult and insecure. Moreover, even now it is the larger land-owners, the really large-scale farmers who work on capitalist lines, and the new vested interests in land, who gain the major advantages of the wartime situation. As I have already pointed out, the government wartime machinery works always in favour of these interests as against the small farmers, the smallholders and the market gardeners, whose profits from the situation are, comparatively speaking, much smaller. These small masters are as much victims of the contemporary economic system as their employees, and circumstances frequently make it difficult for them to be other than mean and grasping.

Any new system of agriculture must find a place for the working farmers, as their knowledge and industry make them an indispensible section of the workers who live from the land.

As farmers, in spite of their skill and specialist knowledge, are economically far below employers in other industries, so agricultural labourers are far below other workers. Before the war, the average wages of an agricultural labourer compared with a builders' labourer was, respectively, 32s. and 52s. 2d. At the commencement of the war, when the Government realized that some concession must be made to such a necessary section of the working class, the wages level was raised to 48s. Farm labourers received more than they had ever earned before. But the increase was largely illusory. Since then the cost of living has risen steadily and in almost every other industry there have been successive increases in wages, but in agriculture the increases demanded by the workers' representatives were deferred time and time again. Negotiations started in April 1941, and after six meetings of the Central Wages Board it was only in November that an increase was granted—to 56s., instead of the very modest £3 asked by the workers. After further agitation it was finally decided, at the end of the year, to grant the full £3 demanded. The gain is, of course, so much Dead Sea fruit, because since the original demand was made, more than six months before, prices have continued to rise to such an extent that the labourer to-day is no better off with £3 than he was at the commencement of the war with 48s.

This record serves to expose the way in which our ruling class regard a body of men who are essential to the community's very life, but who are not sufficiently organized to enforce their wishes. A further example of the methods of the ruling class towards the workers is given by the campaign in certain sections of the yellow press, in which, while lip service was paid to the justice of the claims made by agricultural workers, it was continually stressed that increases in wages would mean increases in the prices of food consumed by industrial workers. Thus it was hoped to exploit the apparently differing interests of two sections of the working class.

In many ways, other than remuneration, the agricultural workers are badly treated. Housing, public utilities and social services are all inadequate. House rents are low, but the cottages are mostly out of date, inconvenient and insanitary and often in a poor state of repair. Damp houses are responsible for a high incidence of rheumatic diseases among the country people. In parts of East Anglia the medieval wattle-and-daub style of architecture still exists in the villages. Council schemes for rehousing have been scanty and unsatisfactory in that the rents of the new houses are considerably more than those of old cottages. A further point of grievance is the survival of the system of tied cottages by which a cottage goes with a job and loss of employment means also loss of a roof over the head of the unfortunate labourer and his family.

Electricity, water supplies, sanitation, medical services and transport are also more absent than otherwise in the country. As an example I will describe the amenities of an East Anglian village in which I lived for a short time and which lies within five miles of the cultured city of Cambridge. There was no electricity supply, and the water supply consisted of two public pumps, from which some of the cottages were removed half a mile or more. The villagers obtained water for washing and all other purposes except cooking and drinking from filthy scummed pits in their gardens. A few favoured cottagers had earth closets, but most had to put up with bucket latrines. Except on Wednesdays and Saturdays, it was necessary to walk 1½ miles to the bus route, and the nearest doctor was five miles away. There are many other places, in all parts of the country, as badly served as this village.

One of the principal reasons for the poor status of the agricultural worker is his lack of organization. The National Union of Agricultural Workers has a membership of less than 10% of the total number of agricultural workers. The farm workers in some areas are organized—by a curious territorial agreement—in the Transport and General Workers' Union. The number of workers affected is probably relatively few, but it is an example of bad organization which must weaken the effectiveness of the men's union activity.

Not only is the number of organized workers small, but they are confined for the most part to limited areas, such as East Anglia, where a certain radicalism exists among the farm workers. Over the larger part of the country the membership of the unions is negligible.

There are a number of reasons for the strong disinclination on the part of many farm workers to join unions. Firstly, as many of the young men were attracted away to work in the towns, a very large proportion of farm workers are elderly men and men who have remained in farming because of their lack of enterprise and initiative. Such men are not good material for unions, and they are very often conservative and suspicious of anything beyond their normal experience in the village.

A second and very important difficulty lies in the fact that instead of being collected in large groups, like factory workers, the farm workers are scattered among many small employers. Few farms employ ten or more men and the majority are worked by a much smaller number of hands, down to the many small farms with only one labourer. It is thus much more difficult to approach the men to obtain the necessary contacts for inducing them to enter the union, and also to maintain their interest once they belong.

Thirdly, there is the motive of fear. The agricultural labourer, in periods of stagnation or decline, is peculiarly liable to victimization. Membership of a union often leads to dismissal, and if the man lives in a tied cottage he may lose his house as well as his job. Furthermore, it is the custom in some parts of the country to black-list unionists, and a man may well find that every farmer in his district will refuse him employment.

Fourthly, the patriarchal tradition persists where there are good employers, and a squire or farmer who treats his men well can induce in them a loyalty that makes them disinclined to take any action they feel detrimental to his interests.

Trade unionism has been a more patent failure in agriculture than in any other industry. This is partly because a form of workers' organization based on wage increases under the present economic system is bound to be ineffective in a declining industry in which it is normally economically impossible for an employer to grant improvements in the conditions of employment, however just or desirable such improvements may be. It is also partly due to the fact that the trade unionists have tried to apply to agriculture a form of unionism based on the large aggregate of workers, the factory, the office and the workshop. To impose such a union on farming is to disregard the pre-industrial character of agricultural employment.

Since it is obvious that, under the present economic system, there can in normal times be no substantial improvements in agricultural wages or conditions of work, it follows that an organization based merely on the struggle for such improvements is useless and redundant. An improvement in the condition of agricultural workers can only come with a permanent expansion of agricultural production in this country, and such an expansion can only occur with the termination of the system of exporting capitalism. Therefore any organization of farm workers must be based on some objective more real and more revolutionary than mere increases in wages. It must aim at a complete change in the social and economic structure of British agriculture, a change that will remove both the injustices and the economic insecurity so intimately associated with the present system.

5

ANARCHISM AND AGRICULTURE

Before I describe the anarchist proposals for agriculture, it is desirable to devote some space to a brief outline of the anarchist social theory.

Anarchism is the doctrine of society without government. It teaches that the major economic and social injustices are intimately associated with the institution of government, which inevitably, in whatever form it takes, creates privilege and a class system, and, even if it may call itself democratic, must base itself on the coercion of the individual, at best to the wishes of the majority, most often to those of the governing class.

Anarchists believe that society should not take the form of a great super-individual body enslaving all its subjects in the interests of the few, but that it should be based on the free co-operation of individual men and women in fulfilment of their common functional and economic needs. In the words of Saint Simon, we believe that

> 'A time will come when the art of governing men will disappear. A new art will take its place, the art of administering things.'

It is in this 'administration of things', in the necessary production and distribution of goods consumed by men, that anarchists see the need for organization, on a voluntary and co-operative basis, among the individuals whose work actually produces the necessities of a civilized life. The functions of the modern state, represented by its paraphernalia of legal codes, bureaucracy, army and police, we consider to be wholly unnecessary in a society where common ownership has ended privilege and social-economic inequalities. Under anarchism every man, once he has fulfilled his contractual economic functions, will be free to live as he likes, provided he does not interfere with the lives of his fellows, and a free people can be relied on to see that the peace is maintained under such circumstances without the need of police or magistrates.

We believe then that the land (like all means of production and the products thereof) should be the property of society held in common, and that only when land has been expropriated can there evolve a satisfactory agricultural system, which will both use the land to its full capacity and ensure to the workers a just and adequate standard of life.

We do not, however, desire the nationalization of the land, as do most of the 'socialist' parties, whether reformist like the Labour Party, semi-revolutionary like the I.L.P., or merely conservative like the Communist Party. We do not desire a Post Office agriculture in which private or trust capitalism will merely be replaced by state capitalism.

We desire that the land shall belong directly to the people, and that it shall be vested in those members of society who are fit and willing to work it, organized in economic federations to provide for society in general the full benefits of an earth freed from exploitation, either of individual capitalists or of the state. Such an organization of agriculture, liberated from the selfish motives of vested interests and from the economic necessities that under capitalism force the ruling class not only to neglect but, in times of peace, actively to restrict the production of food from English soil, we consider to be the most efficient for attaining the dual objective outlined in the beginning of this chapter.

It must be emphasized that such a reconstruction can be successful only as part of a revolutionary reorganization of society on a basis of common ownership and free co-operation in the workers' economic organizations. Without such a syndicalist reorganization it would be impossible for an agricultural system based on workers' control to function effectively, as, in the very unlikely event of a capitalist government leaving it unmolested, the needs of an exporting industrial capitalism, to which under any circumstances like the present it would inevitably remain subservient, would force it into a similar economic position to that of neo-feudal agriculture to-day. It is virtually impossible to establish anarchism in one industry in a country whose present form of society would be in economic and social contradiction.

I do not, however, mean that under the present form of society some progress might not be made towards anarchist organization. Such progress might be represented by the formation of farmers' co-operatives both for selling produce and for buying seeds, fertilizers, plant, etc. It might also be represented by experiments in communal farming, some of which are being evolved in this country at the present time.

But it must be borne in mind that such organizations, whether co-operatives or communities, are dependent on the society in which they exist. Communities live truly on sufference, and a new move towards large-scale agriculture, such as seems a possible outcome of the advance of real capitalism, represented by industrial concerns, banks, etc. into the system of land tenure, might have destructive effect on them as well as on the tenant farmers. Co-operatives have been successful in several countries, notably Denmark, Holland, French Canada and Ireland (in the last two countries under the influence of the Roman Church). But it is noticeable that their success has been greatest in countries whose agricultural industry rests on an export basis. In Denmark co-operatives and kindred voluntary organizations for such purposes as cattle breeding played a great part in the expansion of farming and, as in other countries where they have been successful, benefited their members by reduced costs for seeds, fertilizers, etc., and higher prices for produce, and also helped independent farmers by forcing the privately owned dairies and wholesalers to raise their prices in order to compete successfully with the co-operatives. Thus in Denmark the co-operatives handled 91% of the dairy produce and 86% of the bacon. They purchased on behalf of their members 67% of the feeding stuffs, 40% of seeds and 38% of fertilizers. In Holland a large proportion of the sugar beet, straw board and potato flour factories were operated by co-operatives, which in all cases forced a general rise in prices paid to producers.

In a declining market, however, such as existed in England up to the outbreak of war and is likely to continue afterwards if the capitalist system prevails, the co-operatives would lose much of their value, as the stimulus they gave the Danish farmer to increase his production would in England, under adverse circumstances, become a danger to the farmer by encouraging him to produce more goods than he could sell or, alternatively, so much produce that he would find the prices forced down to an uneconomic level. In any case, farmers' co-operatives would impinge on the Government's policy of marketing boards, and could only become established if the farmers declared a boycott on the government marketing organizations and insisted on trading only through their own co-operatives.

Co-operative or communal experiments in agriculture within a feudal or bourgeois society may, therefore, attain a certain amelioration of conditions for the farming class, but such improvements will be dependent on both economic and political conditions, and can only be regarded as temporary. No stable and permanent social and economic reorganization of agriculture can occur except in a revolutionary society. The real Agricultural Revolution will be part of the Social Revolution.

The methods of struggle in the countryside and the economic organization of the agricultural population will be discussed in the last section of this essay. The remainder of the present chapter will be devoted to an outline of an anarchist agricultural system, for it is necessary to know the nature of our object before we decide on the nature of the struggle we should pursue.

Anarchist agriculture would not be based on growing those crops which would gain most financial profit, not would it be restricted by the needs of an exporting industrial capitalism to maintain a large market for imported food. It would be founded entirely on the exploitation of the soil to its full capacity in order to grow an abundance of the food necessary for the population of this island.

In order that the soil might be cultivated as intensively as possible, it would be divided into comparatively small units, worked by groups holding the land in common, and organized into collectives or syndicates. In general, the syndicate would correspond with the village, and thus the village commune would be revived as a living functional unit.

The village syndicates would embrace not only the farm workers but also those rural workers whose occupations, while not directly agricultural, are necessary to farming, e.g., blacksmiths, bricklayers, wheelwrights, carpenters, mechanics, etc. It would satisfy the common needs of its component groups. Farm machinery would be held in common by the workers in the syndicate, which would arrange the allocation of machinery among the groups. It would also arrange the distribution of seeds, fertilizers, feeding stuffs for cattle, and other products necessary for agricultural work. It would arrange veterinary services and the destruction of pests, and at times, such as harvest, when co-ordinated work was necessary, it would arrange this as well.

The village syndicates would be grouped in district federations, and these again would be united in a national federation. The federations would conduct agricultural research and education. Under anarchism the science of agriculture would cease to be academic, and would become intimately connected with the practice of farming, so that any discoveries that might heighten the productivity of the soil or reduce the effort of cultivation would find general and immediate application.

The federations would maintain close contact with factories and workshops making and repairing agricultural machinery and the chemical factories manufacturing artificial fertilizers. They would arrange with the syndicates of food preparation workers and distributive workers the provision of fresh food to the non-agricultural population and of raw materials for manufactured food. Collection centres for agricultural products, dairies, bacon factories, canning factories and other establishments where simple processing takes place and where close proximity to the growing area is desirable would be embodied in the agricultural syndicates.

The federations would arrange with the distributive syndicates for the provision of non-agricultural goods necessary for the farm workers, and with the appropriate service syndicates for the provision of amenities in the country districts, such as transport and health services, housing, water and electricity supplies, etc.

Anarchist society in general would be regulated on the principle of 'From each according to his ability, to each according to his needs'. The wages system would be superseded by the distribution of goods, and in this distribution no man would be favoured because of his function. A stockman would receive no more than a general labourer. On the other hand, a man with a family—and therefore greater needs—would receive more than a single man with no children.

Administration would be in the hands of the workers. Each farm group would be autonomous so far as its own affairs were concerned, and the assembled members would reach all decisions affecting the work and administration of the farm. The village syndicate would co-ordinate the various groups, and all decisions regarding village matters would be agreed among the members, who would appoint a delegate committee to administer the decisions of the village assembly. This assembly would govern not only the agricultural and economic co-ordination of village life, but also the municipal functions of the present parish councils and arbitration in the event of disputes between members. The village would appoint delegates to the regional federations, which in their turn would appoint delegates to the national federations. No delegate would have power to speak for anything but the decisions of the workers who elected him, and would be subject to recall at any time. He would be elected for a comparatively short period, as would any officials who might be found necessary. Neither delegates nor officials would enjoy a standard of living higher than that of the agricultural workers themselves.

With this form of organization, agriculture, like every other function in an anarchist society, would be administered from the bottom of the pyramid. All decisions would be those of the workers, conveyed by delegates with no 'representative' role and administered by a minimum number of officials elected for short periods and paid at a rate no greater than that of the workers they served. So would be prevented the rise of a powerful bureaucracy and the appearance of a new class basis within the industry.

An agriculture based on a sound economic and social basis would provide for an increase in the country population, and it is likely that there would be a large and increasing shift of population back from the centres of industry into the rural areas. An industrial system not concerned with exportation on an imperialist basis and in which all the scientific means had been used for reducing labour, would release many workers for the land, and with this increase in the working population and a full mechanization of agriculture, a more intensive cultivation could be introduced at the same time as a considerable increase in leisure.

This is an outline of the proposals which anarchists advance for the reorganization of agriculture. And it should be emphasized that our ideas are not based on theory merely, but also on the concrete example of the land workers' collectives in Spain. In July 1936, at the commencement of the Spanish Civil War, revolutionary action was taken by the peasants and workers in many parts of anti-Franco Spain (in particular Catalonia and the part of Aragon which Durutti's columns liberated in the early months of the war) and they carried out large-scale expropriations of land, as well as industrial and transport undertakings. The factories and transport services were managed, with very much increased efficiency, by the syndicates of the workers in their respective industries, and the land was, for the most part, taken over and worked collectively by the peasants, acting in free co-operation. The collectivization was very extensive; in Aragon it is estimated that some 75% of the land was worked by collectives, and in Catalonia the proportion was even higher, in the region of 90%. But, in spite of the widespread nature of the movement for collectivization, it was carried out on an entirely free basis, and no compulsion was attempted by the anarchists to force individual peasants to join the collectives. On the contrary, where peasants elected to remain independent, the collectives assisted them in every possible way, and even allotted to them extra parcels of land to increase their holdings to the size necessary for a reasonable standard of living. Nor were the collectives in any way or at any time due to the actions of the Spanish Republican Government. They were established entirely by the free and spontaneous action of the peasants themselves, and all the government did was unwillingly to recognize the *fait accompli* and issue decrees confirming collectivization. When, later in the war, through the lack of vigilance of the workers and the treachery of Communists and Socialists, the government had become strong on Russian arms and Bank of Spain gold, it began the destruction of the collectives—which was only completed by Franco's armies.

During the period when power remained in the hands of the workers, the condition of agriculture under the collectives was improved, and everywhere the peasant standard of life became higher. There were technical improvements in all types of farming. Selection of seeds, the use of chemical fertilizers and the introduction of farm machinery (often into districts where before it had been unknown) resulted in an increase in the productivity of the land and a simultaneous reduction in the labour necessary for its cultivation. The average increase in wheat yields was approximately 30%, and there were smaller but appreciable increases in the yields of other crops, both cereal and root. Irrigation was greatly extended, and new orchards were planted. It was, however, in stock breeding that the most remarkable results were obtained, and in Aragon the number of cattle and pigs was tripled during a period of eighteen months. Owing to a process of selective elimination of diseased beasts, the stock became healthier, and the extension of cultivation to hitherto unploughed lands produced an ample supply of cattle food.

This increase in agricultural productivity, together with the application of the principle of mutual aid to village affairs, brought about an improvement of the peasant's life. Each person, working according to his ability, received according to his needs of the necessities of life. The community cared for the aged and the unfit, and through the federations of collectives the poorer villages were assisted by the more prosperous, while by arrangement with the health and education syndicates medical services and schools were established everywhere in the rural areas.

Thus, both by the increase of the food production of the country and the amelioration of the conditions of the land workers, the anarchist organization of agriculture in Catalonia and Aragon, carried out by the free and direct action of the workers themselves, proved in practice the value of the type of revolutionary change we propose.

It must be emphasized that, though certain plans can be laid down for the reorganization of the farming industry immediately after the revolution, anarchism does not envisage a static blueprint future for the world. On the contrary, when men have been freed from economic and social oppressions, the evolution of human institutions will probably attain forms we cannot imagine, and therefore, though we can make proposals for a scheme of agricultural organization immediately after the revolution, this must not be regarded as something permanent and therefore dead, but as the basis of further social developments.

6

THE STRUGGLE IN THE COUNTRY

The social and economic struggle in the country is not based on the relatively clearly defined opposition between worker and capitalist that exists in the industrial areas. For, as we have seen, there is in the country a class of farmers who, themselves exploited by the landowning and capitalist classes and their creatures the state and church, are in their turn forced by circumstances to exploit to the utmost the landless labourers they employ.

Thus a deep division exists between farmer and labourer which, before any reorganization of agriculture can take place, must be closed by a realization of the virtual identity of their interests against the exploiters and of the necessity for them to unite in attacking the forces that oppose the rebuilding of farming on a just and prosperous basis.

If effective action is to be taken towards the reorganization of agriculture, it is essential that a measure of unity should be reached between the farmers (and particularly the medium and small farmers) and the wage earners. As I have said before, a body of men who represent a third of the agricultural population and whose practical knowledge of farming is so large, cannot be left out of a collectivized agriculture.

A change in the agricultural system would benefit the smaller farmers as much as the labourers and the consumers. Men whose normal income from a life of endless hard work and financial risk is as low as that of a builders' labourer can lose nothing from an expropriation of the land, which would, indeed, be detrimental only to landowners and the really large farmers. Under collectivization they would be assured a reasonable standard of living without the insecurity of the tenant farmer's life and the crushing exploitation to which he is subjected by landlords, tithe-holders, middlemen, etc. However such men may imagine they are free and independent, they are in reality more enslaved than their labourers, for the labourer has one master where the farmer has many—and all rapacious and merciless. Under anarchism he would be truly free, free of exploitation and assured of a comfortable and masterless life based on the co-operative work of the collective.

As the interests of labourer and working farmer are thus virtually identical, and as the farmer is in any case essential for the success of the new agriculture, whatever methods of struggle we advocate must aim at uniting these two classes in their fight for freedom, and must attempt to avoid antagonism and difference between them.

Thus, while the farm workers should seize every chance of improving their conditions, and of working towards a revolutionary change in the agricultural system, they should always bear in mind the desirability of common action between them and the working farmers. A certain caution will be particularly necessary in the matter of strikes. Where the differences are between the men and the farmers, every attempt should be made to avoid unnecessary strikes and to settle matters by friendly arrangement. Where the differences are on a broader scale and the strike can be used to attack a vested interest or the government itself, the workers should make clear to the working farmers that the strike is not aimed at them, and should endeavour to gain their co-operation in the struggle.

There are forms of direct action which could be used to great effect by a union of farmers and men. In the event of a revolutionary general strike, the farming industry could be of vital importance if it supplied food to the workers' organizations and withheld it from the government. A similar technique could be used under the capitalist régime to force better conditions for the farming industry. Farmers could refuse to sell food at unfair prices, and could also refuse to send food to the government marketing organizations. This latter move could be coupled with the formation of co-operative selling organizations.

Other forms of action in which the small farmers could act in co-operation with the labourers would be rent and tithe strikes. These, by attacking the landlord and the state, would be as effective in a revolutionary manner as the sales boycott. Rent strikes were a favourite method during the land agitations in Ireland, and in England in recent years there were large anti-tithe movements among the farmers. In certain parts of the country where farmers were sold out for non-payment of tithes, the local farmers formed themselves into bands who attended the auctions of distrained cattle and implements and forced the sale of cattle for a few shillings, so that they might be handed back to their owners. (This movement, curiously enough, had the support of certain sections of the capitalist press, notably the Beaverbrook papers.)

But, while it is desirable that the small farmers and the agricultural workers should act together in their fight against the exploiters, the case is different with regard to the large-scale capitalist farmers and nurserymen, and the state organizations employing agricultural labour (e.g. War Agricultural Committees). State organizations stand for the moneyed classes against the poor and moderate men, whether labourer or farmer, and their interference in economic matters which are the concern only of the land workers should be resisted in every way, whether by boycott, by strike or by any other means likely to interfere successfully with their success. As for the large-scale capitalist farmers and nurserymen, these are exploiters in the true sense of the word, in that, while the small farmer works with his labourers, these men merely direct and exploit the labour of others, and, furthermore, use the advantage conferred by their capital resources to employ labour-saving machines and methods which decrease labour costs and enable them to undercut the small growers whose obsolete equipment forces them to ask relatively high prices for their goods.

The large farmers are the enemies of both the small farmers and the labourers. They are true capitalist employers, the vanguard of the new capitalist structure that is likely to govern agriculture in the default of a revolutionary change in the system. With them the small farmers have little in common; much less, indeed, than they have in common with their labourers. There is, therefore, no reason why the employees of the large farmers should not use the strike or their small competitors the boycott, where either form of attack can be used to advantage. The advance of capitalism into the country is to be regarded with the utmost hostility. The new vested interest in land of the financier and industrialists is the beginning of a movement of large-scale capital against the farming class, a movement that might well result in a new wave of enclosures and misery in the countryside. It is therefore necessary that the country people should combat any manifestation of capitalism that appears in their midst.

In order to carry on the social struggle in the country, it is necessary for the land workers to be organized. For this purpose the existing organizations are all inadequate. We do not want a union limited to agricultural labourers, or one, like the National Farmers' Union, which unites the mutually antagonistic sections of small farmers and large farmers but excludes the wage labourers. We do not want a union based on wage increases or price increases under the present system. We do not want a friendly society or a coffin club. We do not want an organization that is tied to the interests of one section of the land workers and that has for its object merely the wresting from the hands of the reluctant ruling class of a few meagre benefits to mitigate the hard and unprofitable life of the country.

The union we desire for farm workers is one great industrial union of the countryside, a union of all the people who gain their living from working the land and not merely from the exploitation of wage earners. This union should include farmers and labourers, smallholders and market gardeners, the village craftsmen whose work is necessary to agriculture and the village women who work in the fields at harvest and potato sowing. It should have for its object not merely minor and temporary mitigations of the present system governing and stultifying the country life, but a revolutionary change in that system which would bring about the rural society of free co-operation and common ownership which we propose. It should aim at a standard of living for land workers which would replace to-day's toil and poverty by to-morrow's leisure and abundance. It should aim at a standard of agricultural production which would give a plenitude of food from our own soil such that no person in the land need want for ample sustenance. It should be a union of countrymen bent on reviving the fertility of the country and restoring agriculture to its position of the major industry in a land from which exploitation and scarcity have been expelled with the system that bred them.

This great union of land workers would build itself around the natural institutions of the country. Its normal unit would be the village, where it would endeavour to arouse again the communal consciousness which existed in the medieval villages and which dwindled almost to nothing with the destruction of the English peasantry. Each village union would be autonomous, and would make and pursue its own decisions without the dictatorial centralism that characterizes and stunts normal trade unionism. The village unions would, however, be joined in county federations for common action and mutual assistance, and the county federations would be united in a national federation. Thus the unions would act as the nuclei of the syndicates or collectives which would administer agriculture under an anarchist system. Like these syndicates, they would employ a minimum of officials, and these for small period and at salaries similar to those of the farm workers. In this way could be avoided the elevation of a corrupt bureaucracy, which would betray the workers as the trade union leaders have done to-day.

While this union would not neglect to carry on, whenever possible, the daily struggle for all classes of the agricultural population, and against landowners, capitalists, state and church, its main object would be the formation of a consciousness among the country people of the identity of their interests against the exploiters and of the necessity for a revolutionary change in the economic and social bases of country life.

Perhaps the most important duty of the unions at the present time would be education, both in the economic and social concepts necessary to produce the consciousness of the need for the true Agricultural Revolution as a prelude to freedom, leisure and plenty, and in the advanced methods of agriculture which would ensure the use of the land to its full capacity.

In furthering this task the unions might establish popular centres of education in the country, similar to the Danish Folk Schools, but with a revolutionary object. Such schools might well become the nuclei from which the new revolutionary urge would spread through the rural areas and arouse the land workers to the necessities of their situation.

To these ends all those interested in the welfare of the country and its people should work. The land of England can never be fertile, nor its workers free and prosperous, until the feudal and capitalist systems are ended and replaced by common ownership and free co-operation in production. In the words of Gerrard Winstanley, the Digger,

> 'True freedom lies where a man receives his nourishment and preservation, and that is in the use of the Earth.'

Lightning Source UK Ltd.
Milton Keynes UK
UKHW022054300123
416216UK00017B/231